I0762841

Blue-Footed Booby

Eastern
Coachwhip

Nutria

Texas Horned Lizard

Pronghorn Antelope

Chimney Swift

Karner Blue
Butterfly

Wild Hog

Striped Bark
Scorpion

WEIRD, WILD, *and* RARE

EXTRAORDINARY ANIMALS *of the* UNITED STATES

WORDS BY
Elizabeth Eakes

PICTURES BY
Bindy James

duopress
an imprint of sourcebooks

TABLE OF CONTENTS

Introduction 3

Extreme Habitats 4

Super Senses 6

What's for Dinner? 8

Let's Look at Legs! 10

Protect Us! Animals at Risk..........................12

Powerful Protection 14

Concealed by Camouflage..........................16

What's Your Wingspan?...............................18

Amazing Adaptations 20

Creepy Creatures...22

Disgusting Diets 25

Picky with Places: Most Specific Habitats..26

Extreme Sizes, from Tiny...to Huge!............28

Brilliant Builders...30

What's That Sound? 32

Vicious Hunters 34

Animals with Extraordinary Abilities 38

Smartest Species..40

First to the Finish Line 42

Observing Animals Safely............................44

How to Help Endangered Animals 45

Glossary...46

Index...48

INTRODUCTION

Have you ever seen a fish without eyes? Or a bird that can do puzzles and count?

WELCOME TO THE UNITED STATES, a country with many different environments. From deserts and plains to mountains and forests, our fifty states have it all—even icy glaciers! And in those environments, you will find some of the weirdest, wildest, and rarest animals on Earth.

Did you know there is a lizard that can shoot blood out of its eyes?

Weird: STRANGE, ODD, OR CRAZY

Or that black widow babies sometimes eat their siblings as they hatch from their eggs!

Wild: FIERCE, SCARY, OR STRONG

Did you know that manatees are very peaceful and intelligent animals? And sadly, they are an endangered species.

RARE: ENDANGERED, VULNERABLE, OR LIVES IN A VERY SPECIFIC PLACE AND IS HARD TO FIND

LOOK FOR THESE SYMBOLS IN THE BOOK!

NT **Near Threatened:** The population is starting to get smaller, and they may become vulnerable or endangered in the future.

V **Vulnerable:** The population is getting smaller by 30–50%, and they are considered a threatened population.

E **Endangered:** The population is getting smaller by 50–70%, and they are at a high risk of becoming extinct.

CE **Critically Endangered:** The population has gotten smaller by 80–90%, and they are at the highest risk of becoming extinct.

In this book, you will learn about all these fascinating animals and many others. We will travel across the United States to discover animals with super senses, disgusting diets, and incredible talents. Each chapter will teach you about a group of animals with special traits or skills. Then, you'll learn how to watch wild animals safely and how to help endangered animals survive. By the end of the book, you'll be an expert on weird, wild, and rare animals!

EXTREME HABITATS

The place where humans live is called a home. Your home might be a house, apartment, or trailer. But animals live in many different **environments**. The place where animals live is called a **habitat**. And some animals have crazy habitats!

Ice Worm

FOUND IN: Alaska, Washington, Oregon

SIZE: 1 cm long; 1 mm wide

DIET: Snow algae and bacteria

LIFE EXPECTANCY: 5–10 years

FUN FACT: Many tiny organisms live in glacial ice, but they are too small for humans to see with our eyes. Ice worms are the only animals in glacial ice that are big enough for us to see.

Ice worms live in **glaciers** and slowly move through blocks of ice. Brrrr! These worms don't have eyes. Instead, they use organs on the surface of their bodies to detect temperature and light.

Muddy Rocksnail 

FOUND IN: Alabama, Kentucky, Tennessee

SIZE: 0.6–1 inch (15–30 mm)

DIET: Decaying plant matter

LIFE EXPECTANCY: 2–6 years

Would you enjoy living in mud? Kentucky's muddy rocksnails do! The snails bury themselves in the mud in rivers and streams. They have a **gill** that allows them to breathe underwater and in the mud!

Horned Puffin

FOUND IN: Alaska, Oregon, California
SIZE: 10–14 inches (25–35 cm) tall
DIET: Fish, plankton, krill
LIFE EXPECTANCY: Up to 20 years

Do you think you could fall asleep on a rock? Horned puffins make their nests on rocky cliffs. When they get hungry, they dive off the cliffs into the water below to catch fish! These birds are called "horned" puffins because they have little spikes above their eyes that look like horns.

BABY NAME: Puffling

FUN FACT: Horned puffins are great swimmers! They "fly" through the water and use their feet to change directions. And they swim deep into the ocean—they can swim 100–250 feet (30–76 meters) underwater!

Coyote

FOUND IN: Every U.S. state except Hawaii
SIZE: Up to 3 feet (1 m) tall
DIET: Rodents, rabbits, deer, plants, **carrion**
LIFE EXPECTANCY: 10–13 years

BABY NAME: Pup

Coyotes can live in MANY different habitats. They can survive in grasslands, forests, swamps, deserts, and even cities! They live with their families in groups called packs. Coyotes eat many different types of plants and animals, which allows them to live in such a variety of habitats.

FUN FACT: Coyotes are fast, hungry, and sometimes...funny! There is a famous cartoon character named Wile E. Coyote who has been entertaining kids since 1949. Wile E. is always on the hunt for a speedy bird named Road Runner, but he never seems to catch him!

SUPER SENSES

ANIMALS WITH SUPERPOWERS: VISION

Mantis Shrimp

FOUND IN: Chesapeake Bay (Maryland), coast of South Carolina
SIZE: 4–12 inches (10–30 cm)
DIET: Worms, squid, fish, clams, snails
LIFE EXPECTANCY: 3–6 years

Human eyes are able to process three channels of color. With these three channels, we can see the colors of the rainbow. But the eyes of the mantis shrimp can process twelve channels! And they can detect ultraviolet light. What do you think the world looks like through the eyes of a mantis shrimp?

Most butterflies have four photoreceptors to help them see. Swallowtail butterflies have fifteen photoreceptors! They have the best vision of all butterfly species.

Eastern Tiger Swallowtail Butterfly

FOUND IN: Eastern U.S., from Florida to Vermont
SIZE: Wingspan: 3–5.5 inches (7.5–14 cm)
DIET: Nectar and plant leaves
LIFE EXPECTANCY: 12 days

Bald Eagle

FOUND IN: Every U.S. state except Hawaii
SIZE: Height: 30 inches (76 cm); Wingspan: 72–84 inches (182–213 cm)
DIET: Fish, rodents, small birds
LIFE EXPECTANCY: 15–20 years

FUN FACT: The bald eagle is the national bird of the United States.

If you notice something that is difficult to see, someone may say, "Wow, you've got an eagle eye!" Take this as a compliment because eagles have amazing vision. Especially bald eagles! They can see SUPER far into the distance—five times farther than humans can see! Bald eagles have a 340-degree field of vision (humans have 180 degrees).

SUPER SENSES

ANIMALS WITH SUPERPOWERS: HEARING

The frequency of sounds is measured with a unit called a hertz. The sounds we hear every day are usually between 2,000 and 5,000 hertz. Animals with a larger hertz range have better hearing because they can hear a wider range of frequencies.

HEARING RANGES (HERTZ)

Chincoteague Pony/ Assateague Horse: 55-33,500 Hertz

FOUND IN: Maryland and Virginia

SIZE: 4.5 feet (1.3 m) tall at the shoulder

DIET: Grasses and salt brushes

LIFE EXPECTANCY: 20–25 years

Bobcat: 65,000 Hertz

FOUND IN: Every U.S. state except Delaware and Hawaii

SIZE: 27 inches (69 cm) tall at the shoulder

DIET: Rabbits, rodents, birds, carrion

LIFE EXPECTANCY: 10–12 years

Greater Wax Moth: 300,000 Hertz

FOUND IN: Illinois, Indiana, Iowa, Kansas

SIZE: 1 inch (2.5 cm) wingspan

DIET: Beeswax, honey, pollen

LIFE EXPECTANCY: 10–12 months

FUN FACT:

Greater wax moths have the best hearing of any insect! Their hearing has **evolved** over time to help the moths avoid hungry bats.

WHAT'S FOR DINNER?

How much can you eat when you're hungry?
These animals have unique diets and big appetites!

Blue-Footed Booby

FOUND IN: California
SIZE: Height: 32–34 inches (81–86 cm); Wingspan: almost 5 feet (1.5 m)
DIET: Fish (anchovies, sardines, mackerel, flying fish, squid)
LIFE EXPECTANCY: 15–17 years

You can tell how hungry a blue-footed booby is based on its feet! The blue feet become brighter when they have recently eaten. Brighter blue means the bird has been eating lots of yummy fish. Pale blue means the booby is likely very hungry and ready for a meal.

Raccoon

FOUND IN: Every U.S. state
SIZE: 2–3.5 feet (0.6–1 m)
DIET: Fruit, nuts, corn, insects, amphibians, mice, rabbits, birds, eggs, fish, crayfish, and more!
LIFE EXPECTANCY: 2–5 years

Raccoons will eat ANYTHING. They eat fruit, bugs, dead animals, and even trash! Because of their diet, they can live in every state.

FUN FACT:

Animals with the biggest appetites in the world

Blue whales: Eat 4 tons (3.6 metric tons) of krill each day!

Elephants: Eat 350 lbs (160 kg) of plants each day!

Tigers: Eat up to 100 lbs (45 kg) of meat in one feeding!

Little Brown Myotis Bat (E)

FOUND IN: Most U.S. states except Hawaii
SIZE: Wingspan: 3–4 inches (7.6–10 cm)
DIET: Insects
LIFE EXPECTANCY: 6–20 years

These little brown bats can eat 1,200 bugs in one hour! They love to munch on flying insects, such as mosquitos, caddisflies, moths, grasshoppers, and beetles, as well as spiders.

American Bison (Buffalo) (NT)

FOUND IN: Wyoming, Montana, South Dakota, and other Western states
SIZE: Up to 10 feet (3 m) long; Up to 6 feet (1.8 m) tall
DIET: Grasses
LIFE EXPECTANCY: Up to 20 years

FUN FACT: Bison are very heavy: they weigh up to 2,200 lbs (998 kg). That's the weight of a small car!

How many hours do you eat every day? American bison spend nine to eleven hours eating each day! They wrap their tongues around pieces of grass and then swallow the grass whole. Then it takes eighty hours for their bodies to digest the grass!

LET'S LOOK AT LEGS!

Zero Legs: Prairie Rattlesnake

FOUND IN: Great Plains region
SIZE: 3–5 feet (1–1.5 m) long
DIET: Small mammals such as ground squirrels, ground nesting birds, mice, rats, rabbits, prairie dogs
LIFE EXPECTANCY: 16–20 years

Even though snakes don't have legs, their bodies can protect them. Rattlesnakes use the end of their tails to make a rattling noise. The tail has segments that shake against each other and make a rattling sound. When predators hear the sound, they know a rattlesnake is nearby. The sound scares predators away!

Two Legs: American Flamingo

FOUND IN: Florida and Georgia
SIZE: 4–5 feet (1.2–1.5 m) tall
DIET: Small crustaceans, mollusks, worms, insects, small fish, algae
LIFE EXPECTANCY: 30–40 years

Flamingos have two legs, but they usually stand on one! Standing on one leg helps keep them warm. These beautiful and silly-looking pink birds are not endangered, but only a small number live in the United States.

FUN FACT:

Flamingos eat with their heads upside down! They turn their heads upside down and hang their bills in the water, then stomp their feet to stir up food from the mud.

Four Legs: One-Toed Amphiuma

FOUND IN: Florida, Georgia, Alabama
SIZE: 8–10 inches (20–25 cm)
DIET: Small aquatic invertebrates such as crayfish, insect and fish larvae
LIFE EXPECTANCY: 13–19 years

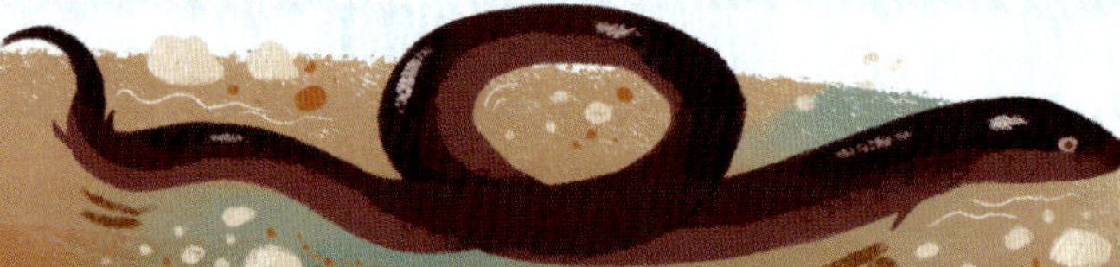

One-toed amphiumas look like eels, but they are actually salamanders. Their tiny legs help them move through the shallow, muddy water where they live. These odd-looking salamanders have only one toe on each foot!

Six Legs: Honeybee

FOUND IN: Every U.S. state
SIZE: 0.5 inch (1.2 cm)
DIET: Nectar and pollen
LIFE EXPECTANCY: Workers: 5 to 6 weeks; Queen: Up to 2 years

Honeybees are dancers! To tell other bees where they can find flowers, honeybees do a special dance called a waggle dance.

FUN FACT:
North Dakota produces the most honey out of all the states in the United States! In 2022, the bees in North Dakota produced 31 million pounds (14 million kg) of honey!

Can you tell how this spider got its name? This smiley spider can only be found on the islands of Hawaii.

Eight Legs: Happy Face Spider 

FOUND IN: Hawaii
SIZE: 0.2 inch (0.5 cm)
DIET: Small insects
LIFE EXPECTANCY: Up to 2 years

Forty Legs: Sonoran Desert Centipede

FOUND IN: Texas, Arizona, California, New Mexico
SIZE: 4–7 inches (10–18 cm)
DIET: Insects, spiders, small vertebrates
LIFE EXPECTANCY: Unknown, estimated to be 5 years

NOT-SO-FUN FACT:
These centipedes can sting! Their stings are painful and venomous.

Sonoran Desert centipedes are nocturnal. That means they spend daytime underground or hiding under rocks and come out at night to hunt.

The Los Angeles thread millipede was discovered in California in 2018. This leggy creature burrows underground in small spaces.

486 Legs: Los Angeles Thread Millipede

FOUND IN: California
SIZE: 1 inch (2.5 cm)
DIET: Decaying plants
LIFE EXPECTANCY: 7–10 years

PROTECT US! ANIMALS AT RISK

Extinction is when a species of animal no longer exists on Earth. **Endangered** means a species of animal is at risk of going extinct. Animals become endangered when their **populations** are growing smaller. These species are endangered, but hopefully, with our help, their populations will grow!

Ivory-Billed Woodpecker 

FOUND IN: Southeastern United States
SIZE: 18–20 inches (45.7–50.8 cm)
DIET: Insects, especially beetle larvae
LIFE EXPECTANCY: Unknown, estimated to be 10–20 years
NUMBER LEFT IN THE WILD: Unknown, possibly zero

The last time an ivory-billed woodpecker was officially seen in the wild was more than eighty years ago! But some people claimed that they saw this red-headed bird in the early 2000s in Arkansas.

Manatee 

FOUND IN: Florida, southeastern United States in coastal areas and rivers
SIZE: Length: 9–10 feet (2.7–3 m); Weight: 800–3,000 lbs (363–1,361 kg)
DIET: Aquatic plants
LIFE EXPECTANCY: 40–60 years
NUMBER LEFT IN THE WILD: 13,000

Manatees have no predators! They are very peaceful and intelligent animals. They are also known as sea cows because they spend many hours a day grazing on seagrasses.

FUN FACT: Manatees are playful swimmers: they can swim upside down, roll, and do somersaults underwater.

This flying squirrel doesn't really fly, but it does glide through the air! It uses stretchy skin called patagium as wings to help glide from tree to tree.

Carolina Northern Flying Squirrel (E)

FOUND IN: South Carolina, Tennessee, Virginia, West Virginia
SIZE: Length: 10–12 inches (25.4–30.4 cm); Weight: 3–6 ounces (85–170 g)
DIET: Fungi, seeds, nuts, insects
LIFE EXPECTANCY: 3–5 years
NUMBER LEFT IN THE WILD: Unknown

Gopher Tortoise (E)

FOUND IN: Southeastern United States
SIZE: 9–15 inches (23–38 cm)
DIET: Grasses, legumes, vegetation
LIFE EXPECTANCY: 40–60 years
NUMBER LEFT IN THE WILD: 700,000

Gopher tortoises love to dig tunnels in the ground, called **burrows**. More than 300 different types of animals live in these burrows.

FUN FACT:

What's the difference between a tortoise and a turtle?

TORTOISE	TURTLE
Eats only plants	Eats plants and animals
Spends most of its time on land	Spends most of its time in or near water
Can live up to 200 years	Lives for 5–50 years
Dome-shaped shell	Flatter, smoother shell

POWERFUL PROTECTION

Many animals have predators. To avoid being eaten, some animals have developed powerful ways to protect themselves.

Virginia Opossum

FOUND: East of the Rocky Mountains and along the West Coast

SIZE: 16-inch (40.6-cm) body and 11-inch (28-cm) tail

DIET: Small mammals, birds, insects, worms, plants, fruits, seeds

LIFE EXPECTANCY: 2 years

FUN FACT:

Virginia opossums are the only **native species** of marsupials in the United States. Native animals are naturally found in a certain environment and spread there without human actions.

When opossums are in danger, they pretend to be dead! Because of this, when someone is pretending to be dead, they are said to be "playing possum."

Nutria

FOUND IN: Gulf Coast states, southeastern United States, Atlantic coast, Pacific Northwest

SIZE: 1–2 feet (30–61 cm)

DIET: Aquatic plants and roots

LIFE EXPECTANCY: 6–8 years

Nutrias defend their burrows aggressively and will bite and scratch humans or pets if they feel threatened.

NOT-SO-FUN FACT:

Nutrias were brought to the U.S. in 1889 so people could use their fur to make clothing. Once released into the wild, they became an **invasive species**. Invasive species are not native to an area and can cause problems in the ecosystem. Nutrias have spread across the U.S. and destroy native plants and wetlands.

Saddleback Caterpillar

FOUND IN: Eastern United States
SIZE: 0.5–1 inch (1.2–2.5 cm)
DIET: Plant leaves
LIFE EXPECTANCY: 4–5 months as a caterpillar, 1–2 weeks as a moth

If you see a saddleback caterpillar, be careful! These spiky insects are covered with spines that are connected to poison glands. If you touch the tips of these spines, they will cause a painful sting. After a few months, the saddleback caterpillar enters a cocoon and emerges as a moth.

Texas Horned Lizard

FOUND IN: Southern-central United States
SIZE: 3–5 inches (7.6–12.7 cm)
DIET: Ants and other small insects
LIFE EXPECTANCY: Up to 5 years

This lizard shoots blood from its eyes to protect itself from other animals! To shoot the blood, the lizard increases pressure in its head, causing blood vessels around its eyes to burst. The blood mixes with a chemical that tastes disgusting to wolves and coyotes.

Magpie

FOUND IN: Northwestern United States
SIZE: 17–21 inches (43–53 cm) long
DIET: Insects, slugs, snails, spiders, fish, reptiles, amphibians, birds, eggs, carrion, seeds, ticks
LIFE EXPECTANCY: 4–6 years

These birds are very protective of their nest. When they have babies in their nest, magpies will fly down and attack people or pets who walk past!

CONCEALED BY CAMOUFLAGE

Camouflage is a strategy that animals use as a disguise. Blending in with their environment protects animals from predators. These animals have special camouflage to help them hide in their habitats!

Arctic Fox

FOUND IN: Alaska

SIZE: 3–4 feet (91–122 cm)

DIET: Small mammals such as lemmings and tundra voles, as well as seabirds, berries, and eggs

LIFE EXPECTANCY: 3–4 years

These foxes can change color! In summer, they grow brown fur to blend in with the ground. Then, in winter, they grow white fur to blend in with the snow.

Stick Bug (Northwestern Walkingstick)

FOUND IN: Eastern, central, and southern United States

SIZE: 2–4 inches (5–10 cm)

DIET: Tree leaves (mostly oaks and hazelnut)

LIFE EXPECTANCY: Up to 2 years

You can probably guess how these bugs were named...they look exactly like sticks! They also sway their bodies side to side to **mimic** branches moving in the breeze. This sneaky disguise helps the bugs hide from hungry birds.

FUN FACT:

Stick bugs can regrow their legs! If a predator grabs them by the leg, they use a special muscle to break off the leg. After they shed their skin, it will grow back.

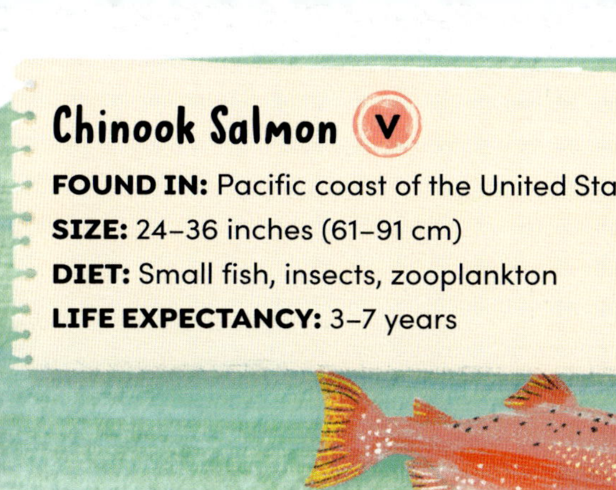

Chinook Salmon (V)

FOUND IN: Pacific coast of the United States
SIZE: 24–36 inches (61–91 cm)
DIET: Small fish, insects, zooplankton
LIFE EXPECTANCY: 3–7 years

FUN FACT:
Female Chinook salmon can carry more than 4,000 eggs at once!

When they are young, Chinook salmon have marks on their sides to help them blend into the freshwater streams where they were born. Once they get older, they migrate to the ocean and their bodies become darker on top and lighter on the bottom to camouflage them in the open sea.

Desert Horned lizard

FOUND IN: Southwestern United States
SIZE: 2.5–5 inches (6.3–12.7 cm)
DIET: Ants and other small insects
LIFE EXPECTANCY: Up to 5 years

Desert horned lizards can change the color of their skin to match the environment! They use this skill to blend in with the sand, rocks, and plants around them.

WHAT'S YOUR WINGSPAN?

Wingspan is the measurement of wings from one tip to the other. Birds are measured by their wingspan. A human's measurement from finger to finger is called an arm span because we don't have wings!

This is the smallest owl in the world! Elf owls live inside cacti and are protected by the spiky needles of the cacti.

Elf Owl

FOUND IN: Southwestern United States

SIZE: Length: 4.5–5.75 inches (11.4–14.6 cm); Wingspan: 9–12 inches (22.9–30.5 cm)

DIET: Insects, small mammals, other small invertebrates

LIFE EXPECTANCY: 3–6 years

Calliope Hummingbird

FOUND IN: Western United States

SIZE: Length: 2.75–3.25 inches (7–8.2 cm); Wingspan: 4 inches (10–11 cm)

DIET: Nectar from flowers

LIFE EXPECTANCY: 3–5 years

The calliope hummingbird is the smallest bird in North America. It weighs less than a nickel! It has a long, forked tongue that is used to lick nectar from flowers.

FUN FACT:

Hummingbirds flap their wings faster than any other type of bird. They can flap their wings seventy times in one second!

California Condor CE

FOUND IN: California, Arizona, Utah
SIZE: Wingspan: 9–10.5 feet (2.7–3.2 m)
DIET: Carrion
LIFE EXPECTANCY: 60–70 years

This bald-headed bird has the largest wings of any bird in the United States. California condors are so big that when they fly high, people sometimes think they are airplanes!

Trumpeter Swan

FOUND IN: Northern United States
SIZE: Length: 55–62 inches (140–157 cm); Wingspan: 6–8 feet (1.8–2.4 m)
DIET: Aquatic plants, grains, small invertebrates
LIFE EXPECTANCY: 10–25 years

These huge birds are the heaviest flying birds in North America. They keep their eggs warm by covering them with their big webbed feet.

FUN FACT:
After hatching from their eggs, baby trumpeter swans don't stay in the nest very long: twenty-four hours after hatching, they can leave the nest to swim and feed.

AMAZING ADAPTATIONS

An *adaptation* is when a body part or behavior helps an animal survive. These animals have unique and amazing adaptations.

Red Hills Salamander (E)

FOUND IN: Alabama, Florida

SIZE: 4–8 inches (10–20.3 cm)

DIET: Insects, spiders, other small invertebrates

LIFE EXPECTANCY: 10–11 years

Red Hills salamanders do not have lungs—they breathe through their skin! Breathing through their skin allows them to live in moist soil.

FUN FACT: The Red Hills salamander is the official state amphibian of Alabama.

Eastern Spadefoot Toad

FOUND IN: The East Coast from Massachusetts to Florida, and from the Mississippi Valley to Tennessee

SIZE: 2–3 inches (5–7.6 cm)

DIET: Beetles, crickets, caterpillars, spiders, snails

LIFE EXPECTANCY: 5–9 years

Eastern spadefoot toads have adapted to survive in dry and unpredictable environments. These toads can burrow underground for eight to ten months if the weather is dry. When they feel the vibrations of rain, they come out from their burrows!

Gray Tree Frog

FOUND IN: Eastern United States

SIZE: 1.25–2 inches (3.1–5 cm)

DIET: Insects

LIFE EXPECTANCY: 7–9 years

Do you like frog-flavored ice pops? This little frog can freeze itself during winter. Gray tree frogs freeze to survive cold temperatures and then thaw out when the weather warms up!

Black-Tailed Jackrabbit

FOUND IN: Western United States

SIZE: 18–25 inches (45.7–63.5 cm)

DIET: Grass and vegetation

LIFE EXPECTANCY: 1–5 years

Look at those giant ears! Black-tailed jack-rabbits have super long ears and super great hearing. Their ears also help keep their body temperature comfortable.

FUN FACT:

Jackrabbits aren't rabbits; they are part of a species called hares! Hares are larger than rabbits and have longer ears.

CREEPY CREATURES

Star-Nosed Mole

FOUND IN: Eastern United States
SIZE: 6–7 inches (15.2–17.7 cm)
DIET: Earthworms, insects, other small invertebrates
LIFE EXPECTANCY: 2–3 years

FUN FACT:
The star-nosed mole can smell underwater! It does this by exhaling bubbles onto objects and then inhaling the bubbles, bringing the scents back into its nose.

Star-nosed moles have a very weird face, with twenty-two tendrils that help the mole move through its environment. Thanks to their star-shaped face, these moles are the fastest-eating mammals in the world! Their special nose helps them find food. They can determine whether an object is food in only 8 milliseconds! Most animals require about .23 of a second (227 milliseconds) to identify and eat food. Star-nosed moles only need .12 of a second!

Alligator Gar

FOUND IN: Mississippi River Valley
SIZE: Up to 10 feet (3 m) long
DIET: Fish, crustaceans, small mammals
LIFE EXPECTANCY: 26–45 years

This species of fish is named after the alligator because of the rows of sharp teeth in its long mouth. It is one of the largest freshwater fish in the world!

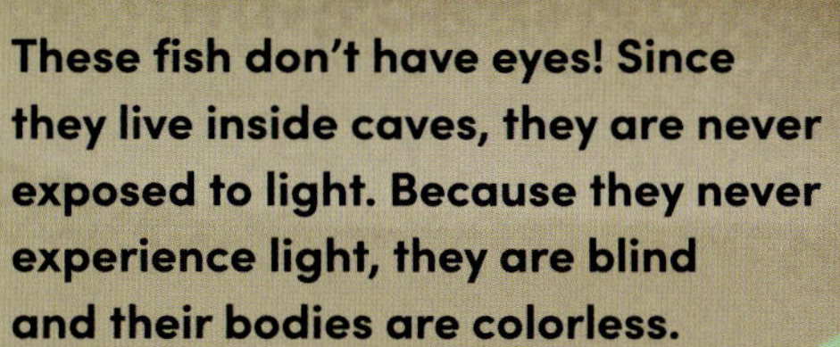

These fish don't have eyes! Since they live inside caves, they are never exposed to light. Because they never experience light, they are blind and their bodies are colorless.

Ozark Cave Fish 

FOUND IN: Arkansas, Oklahoma, Missouri
SIZE: 2–4 inches (5–10 cm)
DIET: Aquatic invertebrates, algae, bat poop
LIFE EXPECTANCY: 10–15 years

Gila Monster

FOUND IN: Arizona, New Mexico
SIZE: 18–24 inches (45.7–61 cm)
DIET: Small mammals, birds, eggs, reptiles
LIFE EXPECTANCY: 20–30 years

These creepy monsters are venomous and can climb trees. They also have an amazing sense of smell; they can smell eggs that are buried 6 inches (15 cm) deep!

FUN FACT:

When trying to protect itself, a Gila monster can bite onto a predator and hold on for up to ten minutes! While biting, it chews and releases venom.

MORE CREEPY CREATURES!

Geoduck

FOUND IN: Coastal areas of the Pacific Northwest
SIZE: Shell is up to 14 inches (35.5 cm) long
DIET: Phytoplankton
LIFE EXPECTANCY: 100–160 years

FUN FACT: Geoducks can live for more than a hundred years!

Geoducks bury themselves deep into sand, and they have very long necks that they use to breathe. They are also known as elephant clams because their necks look like the trunks of elephants.

Wild Hog

FOUND IN: Southern United States
SIZE: Weighs up to 300 lbs (136 kg); Up to 5 feet (1.5 m) long
DIET: Roots, insects, small mammals
LIFE EXPECTANCY: 4–8 years

Wild hogs are not able to sweat. On hot days, they stay near water or lie in mud to keep cool. Wild hogs are also very fast. They can run up to 30 miles (48 km) per hour, which is as fast as a car!

DISGUSTING DIETS

Did you know some animals eat poop? These animals are called scavengers. Scavengers eat dead animals, dead plants, or animal waste.

Tumblebug (Dung Beetle)

FOUND: Throughout the continental United States
SIZE: 0.2–1.5 inches (0.5–3.8 cm)
DIET: Dung
LIFE EXPECTANCY: 1–3 years

Dung means...poop. Tumblebugs are called dung beetles because they eat poop from other animals. They also roll poop into balls! These strange beetles can also use the sun and moon to find their way across land.

Striped Bark Scorpion

FOUND IN: Texas, Oklahoma, Kansas, New Mexico, Colorado, Missouri, Louisiana
SIZE: 2–3 inches (5–7.6 cm)
DIET: Insects
LIFE EXPECTANCY: 3–5 years

These creepy crawlers are named after bark because they love rotting wood and logs. Eww!

FUN FACT:
There are more than 1,500 scorpion species in the world, and they live on every continent except Antarctica.

American Burying Beetle (CE)

FOUND IN: Eastern United States
SIZE: 1.5–2 inches (3.8–5 cm)
DIET: Carrion
LIFE EXPECTANCY: 1–2 years

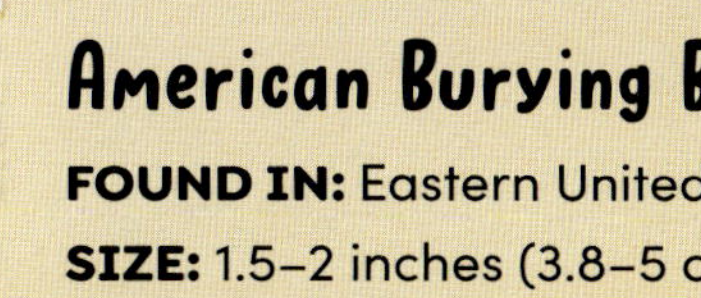

These beetles eat the bodies of dead animals! They also bury the bodies of dead animals to save for later. Leftovers, anyone?

PICKY WITH PLACES: MOST SPECIFIC HABITATS

Squirrel Chimney Cave Shrimp (CE)

FOUND IN: A sinkhole near Gainesville, Florida

SIZE: Up to 1 inch (2.5 cm)

DIET: Organic waste

LIFE EXPECTANCY: Unknown

This shrimp is a tiny animal that has been found in only one sinkhole in Florida. It is so rare that scientists don't know much about it yet!

Karner Blue Butterfly 

FOUND IN: Eastern United States

SIZE: Wingspan: 0.75–1.25 inches (1.9–3.1 cm)

DIET: Nectar from flowers

LIFE EXPECTANCY: 1–2 weeks as a butterfly

This beautiful butterfly needs the lupine plant to survive. However, the lupine plant needs a habitat called oak savanna, which includes a mix of sun and shade. Lupine cannot survive in thick, shaded forests. The United States has less and less oak savanna habitats each year, so lupine plants and the Karner blue butterfly have become very rare.

FUN FACT:

Did you know that butterflies taste with their feet? They have special taste receptors on their feet. These receptors let them know if they have landed on something good to eat.

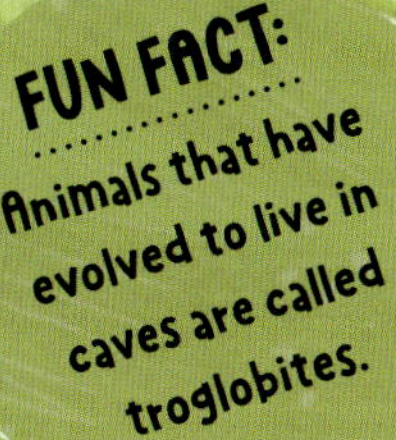

Devils Hole Pupfish 

FOUND IN: Nevada

SIZE: 1.5 inches (3.8 cm)

DIET: Algae and other invertebrates

LIFE EXPECTANCY: Unknown

This tiny fish is bright blue and swims around in a playful way. It swims so playfully that scientists thought it looked like a puppy, so they named it pupfish! These little fish puppies can only be found in the Devils Hole cavern in Nevada, where the water is very hot.

Seneca White Deer

FOUND IN: New York

SIZE: Up to 3.5 feet (1 m) tall

DIET: Grasses and vegetation

LIFE EXPECTANCY: 8–10 years

BABY NAME: Fawn

This rare deer lives only in an enclosed area in New York State. It is the largest herd of white deer in the world.

EXTREME SIZES,
FROM TINY...

Tardigrade

FOUND IN: Every U.S. state
SIZE: 0.5 mm–1.5 mm
DIET: Plant cells, algae, small invertebrates
LIFE EXPECTANCY: 3 months to 2.5 years

Also known as water bears, these animals are so small, it's hard to see them without a microscope. They can enter dormant states and live for up to thirty years without food or water!

FUN FACT: Tardigrades can survive in freezing and boiling temperatures!

American pygmy shrews are the smallest mammals in the United States. Pygmy shrews must eat every fifteen to thirty minutes; if they go an hour without food, they'll die. Because of this, they sleep for only a few minutes at a time.

American Pygmy Shrew

FOUND IN: Northern states, Great Lakes region, and the Appalachians
SIZE: 1.6–2.2 inches (4–5.5 cm)
DIET: Insects, spiders, other small invertebrates
LIFE EXPECTANCY: 1–2 years

...TO HUGE!

Texas Longhorn

FOUND IN: Texas, Oklahoma, Nebraska

SIZE: Length: 6–9 feet (1.8–2.7 m); Weight: 1,000–2,500 lbs (454–1,134 kg)

DIET: Grasses and vegetation

LIFE EXPECTANCY: 15–20 years

The Texas longhorn holds the world record for longest cattle horns: 127.4 inches, which is over 10 feet (3 m) long!

FUN FACT: Even though this animal is named after Texas, many live in Oklahoma and Nebraska.

Kodiak Bear

FOUND IN: Alaska

SIZE: 7–10 feet (2.1–3 m) tall

DIET: Fish, berries, other vegetation

LIFE EXPECTANCY: 20–30 years

Kodiak bears are the largest bears in the world! They live only on certain islands in Alaska (Kodiak Archipelago), and they can grow to be ten feet tall. That's as tall as a basketball hoop!

BRILLIANT BUILDERS

Ruby-Throated Hummingbird

FOUND IN: Eastern United States

SIZE: 2–5 inches (5–12.7 cm)

DIET: Nectar from flowers

LIFE EXPECTANCY: 3–5 years

Hummingbirds like the ruby-throated hummingbird have discovered how to build their homes with creative materials. These tiny birds use plants, moss, and spider silk to create soft and flexible nests. In the fall, they migrate to Central America to stay warm for the winter.

Northern Idaho Ground Squirrel (E)

FOUND IN: Idaho

SIZE: 7–9 inches (17.7–22.8 cm)

DIET: Seeds, nuts, fruit, insects

LIFE EXPECTANCY: 3–6 years

This adorable ground squirrel digs three different types of burrows: one for babies, one for its home, and one for hibernating through winter! Unfortunately, the northern Idaho ground squirrel is endangered. But, thanks to people planting more trees in Idaho, the ground squirrel population has started to grow.

FUN FACT: Squirrels' teeth never stop growing! To prevent their teeth from getting too long, squirrels must gnaw (quickly chew) on bark or wood.

Put on your hard hat and grab your hammer because these beavers can build! Beavers use sticks, grass, and mud to create dams and lodges where they can live. But if you want to see the inside of a beaver's lodge, you'll need to hold your breath; their lodges can only be entered from underwater.

American Beaver

FOUND IN: Most states except the desert regions of the Southwest
SIZE: 2.5–4 feet (0.7–1.2 m)
DIET: Bark, twigs, aquatic plants
LIFE EXPECTANCY: 10–15 years

FUN FACT: Beavers have orange teeth! The coating of their teeth is full of iron, which makes their teeth orange.

Chimney Swift

FOUND IN: Eastern United States
SIZE: Length: 4.5–5 inches (11.4–12.7 cm); Wingspan: 12 inches (30 cm)
DIET: Insects
LIFE EXPECTANCY: 4–6 years

This bird uses "glue" made of spit! It uses its saliva to stick its nest to the insides of chimneys, sheds, or barns.

WHAT'S THAT SOUND?

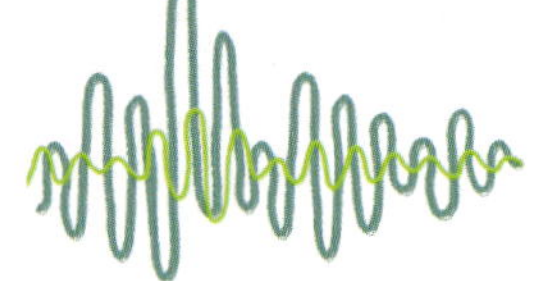

New Jersey Chorus Frog

FOUND IN: New Jersey, Pennsylvania, eastern Maryland, Delaware, Virginia
SIZE: 0.75–1.5 inches (1.9–3.8 cm)
DIET: Flies, springtails, spiders, snails, ants
LIFE EXPECTANCY: 1–3 years

FUN FACT: Female chorus frogs can lay up to 1,500 eggs!

This little frog makes a very strange sound. Male chorus frogs make a clicking sound to attract a mate during the spring.

Attwater's Prairie Chicken (V)

FOUND IN: Texas
SIZE: 17 inches (43 cm) long; Wingspan: 28 inches (71 cm)
DIET: Small green leaves, seeds, insects
LIFE EXPECTANCY: 2–3 years

This bird has big yellow cheeks for making a BOOMING sound. Male prairie chickens dance, stomp their feet, and "boom" to get the attention of the female birds.

American Bittern

FOUND IN: Northern U.S. during summer, migrates to southern U.S. in winter
SIZE: Length: 23–34 inches (58–86 cm); Wingspan: 45–50 inches (114–127 cm)
DIET: Fish, amphibians, insects, small mammals
LIFE EXPECTANCY: 6–10 years

American bitterns live in marshes, bogs, shallow lakes, and ponds. They make a loud and odd call and throw their heads up and forward while calling. These birds are very stealthy and hard to find because they blend into the tall grasses and reeds of their habitat. You may not see them, but you can hear them!

American Red Fox

FOUND IN: Every U.S. state except Hawaii
SIZE: 18–36 inches (45.7–91.4 cm) long; 14–26 inches (35.5–66 cm) tall
DIET: Small mammals, birds, fruit, insects
LIFE EXPECTANCY: 2–4 years

BABY NAME: Kit

American red foxes can make over twenty different sounds. One of their sounds is like a human screaming! They also bark, squeal, and purr.

FUN FACT:

Foxes have whiskers on their faces...and their legs! The whiskers help them **navigate** their environment, especially at night.

VICIOUS HUNTERS

FOUND IN: Louisiana, Texas
SIZE: 4–6 feet (1.2–1.8 m) long
DIET: Rodents, birds, eggs
LIFE EXPECTANCY: 20 years

This is one of the rarest snakes in North America. It lives in tunnels created by gophers and slithers through the tunnels to find its next gopher meal!

Wolverine

FOUND IN: Northern United States
SIZE: 26–34 inches (66–86.3 cm)
DIET: Small mammals, birds, carrion
LIFE EXPECTANCY: 7–12 years

FUN FACT:

Wolverines are so powerful that they have a superhero named after them! In 1974, Wolverine was introduced as a superhero in comic books. Many years later, he became a popular character in TV shows and movies. Wolverine is a vicious fighter with strong metal claws.

The wolverine is a fierce carrion-eater, which means it often eats dead animals. It even has special teeth to help it eat frozen meat!

Fishing Spider

FOUND IN: Eastern United States

SIZE: Body: 0.5–1 inch (1.2–2.5 cm) long; Legs: 3 inches (7.6 cm) long

DIET: Insects, small fish, tadpoles

LIFE EXPECTANCY: 1–2 years

This spider swims to catch its next meal and can stay underwater for thirty minutes!

Grizzly Bear

FOUND IN: Western United States and Alaska

SIZE: 6–9 feet (1.8–2.7 m) tall

DIET: Berries, roots, fish, small mammals, elk, moose, deer

LIFE EXPECTANCY: 20–25 years

Grizzly bears are very aggressive, fierce hunters! They use their claws to dig underground for prey and to catch fish from rivers.

FUN FACT:

Grizzly bears can eat up to 90 lbs (40 kg) of food in one day.

MORE VICIOUS HUNTERS!

Timber Rattlesnake (E)

FOUND IN: Eastern United States
SIZE: Up to 5 feet (1.5 m) long
DIET: Rodents, birds, small mammals
LIFE EXPECTANCY: 10–20 years

The timber rattlesnake is one of the most venomous snakes in the United States and can grow to be five feet long. Luckily, it is not aggressive toward humans!

Fisher

FOUND IN: Northern United States
SIZE: 24–30 inches (61–76 cm)
DIET: Small mammals, birds, carrion
LIFE EXPECTANCY: 7–10 years

The fisher is one of the only mammals that can climb down a tree head-first. It is also one of the only animals that can hunt (and eat!) porcupines.

FUN FACT: Even though fishers are fierce predators, they are very shy toward humans.

Yellow Jackets

FOUND IN: Every U.S. state
SIZE: 0.4–0.6 inch (1–1.5 cm)
DIET: Insects, fruit, carrion
LIFE EXPECTANCY: 1 year

Yellow jackets are active and aggressive at the end of summer, which is when their populations are the largest. Their stingers can sting multiple times.

Black Widow

FOUND IN: Southern United States
SIZE: 0.5 inch (1.2 cm)
DIET: Insects and spiders
LIFE EXPECTANCY: Females: 1–3 years; Males: 1–5 months

Female black widows have a venomous bite. Black widow babies sometimes eat their siblings as they hatch from their eggs!

NOT-SO-FUN FACT:

As Earth's climate becomes warmer, black widows are becoming able to survive in northern habitats. Climate change has expanded their habitat range farther north, into the northeastern United States and Canada.

ANIMALS WITH EXTRAORDINARY ABILITIES

Southern Flying Squirrel

FOUND IN: Eastern United States

SIZE: 8–10 inches (20.3–25.4 cm)

DIET: Seeds, nuts, fruit, insects

LIFE EXPECTANCY: 4–6 years

These flying squirrels usually glide up to 59 feet (18 m). But the longest glide ever recorded was 262 feet (80 m). That's almost the length of a football field!

Nine-Banded Armadillo

FOUND IN: Southeastern states, with their range increasing farther north each year

SIZE: 15–17 inches (38–43 cm)

DIET: Insects and plants

LIFE EXPECTANCY: 7–20 years

These hard-shelled mammals can cross rivers! They inflate and float across, or they hold their breath and walk along the bottom of the river. They can hold their breath for up to six minutes!

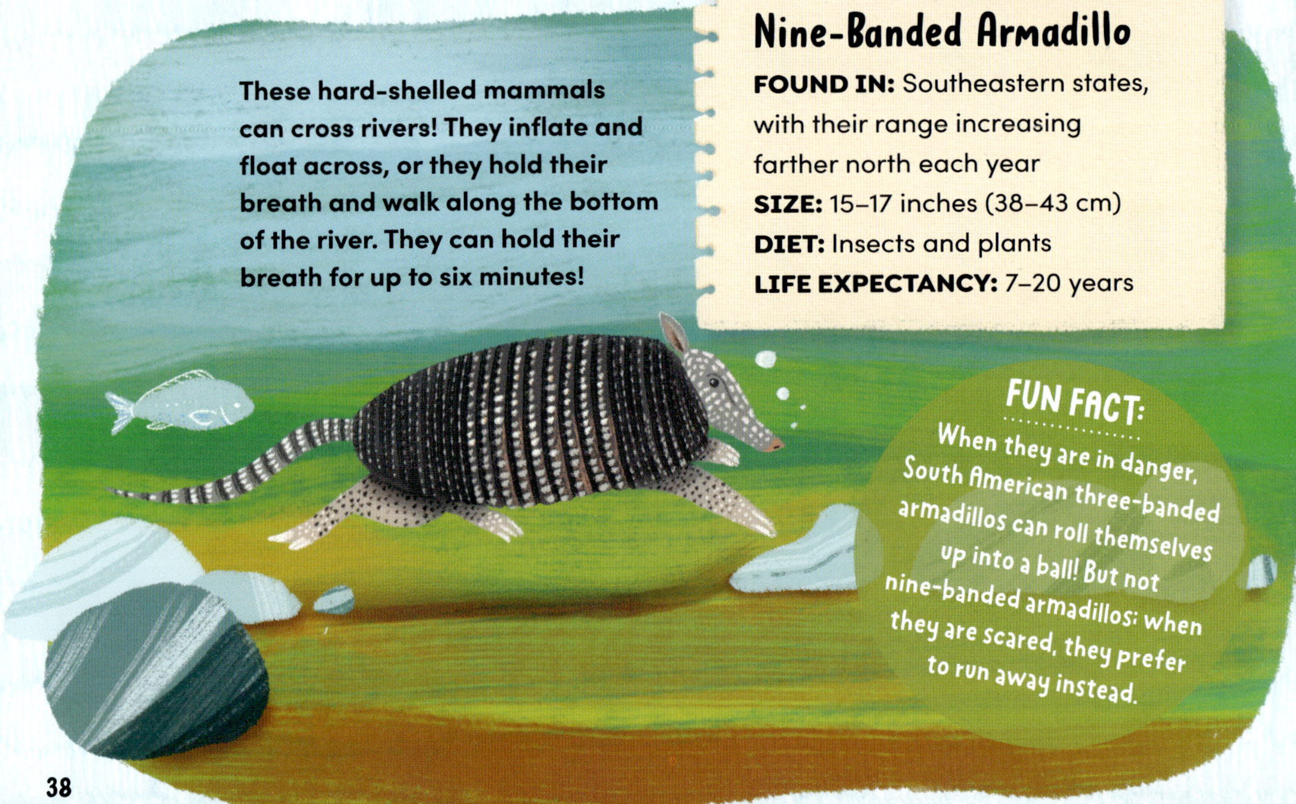

FUN FACT:

When they are in danger, South American three-banded armadillos can roll themselves up into a ball! But not nine-banded armadillos: when they are scared, they prefer to run away instead.

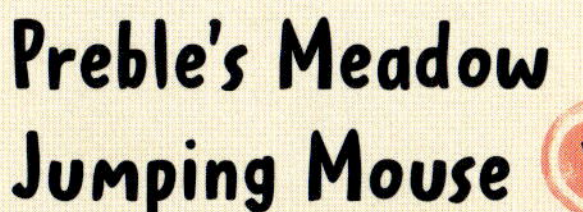

Preble's Meadow Jumping Mouse (V)

FOUND IN: Colorado, Wyoming
SIZE: 7–10 inches (17.7–25.4 cm)
DIET: Seeds, fruit, insects
LIFE EXPECTANCY: 1–2 years

FUN FACT:
Mice are very flexible and can fit into a hole smaller than the width of a pencil.

These tiny mice can use their tails to jump 3 feet (1 m) or more in the air! When they need to avoid predators, Preble's meadow jumping mice spring into the air, using their tails to guide the direction of their jump.

North American Porcupine

FOUND IN: Every region of the United States except the Southeast
SIZE: 25–36 inches (63.5–91.4 cm)
DIET: Tree bark, twigs, leaves
LIFE EXPECTANCY: 5–7 years

Did you know these spiky animals can climb trees? Once they are up in the tree, they go to sleep!

SMARTEST SPECIES

American Crow

FOUND IN: Every U.S. state except Hawaii
SIZE: 16–21 inches (40.6–53.3 cm)
DIET: Seeds, fruit, insects, carrion
LIFE EXPECTANCY: 7–8 years

American crows are very intelligent and have great memories! They can recognize themselves in their reflection and can remember a human for years. They can also count, solve puzzles, and use simple tools to solve problems.

Sea Otter 

FOUND IN: Coastal areas of the Pacific
SIZE: 4–5 feet (1.2–1.5 m)
DIET: Crabs, clams, sea urchins, marine invertebrates
LIFE EXPECTANCY: 10–15 years

Sea otters are very clever. They use rocks as tools to crack open the shells of their prey. They also hold hands while sleeping to prevent them from floating away from each other! This strategy is called rafting.

FUN FACT:
Otters have the thickest fur of any animal. Their thick fur keeps them warm in cold water.

Naked Goby

FOUND IN: Great Lakes region

SIZE: Up to 6 inches (15.2 cm)

DIET: Insects, crustaceans, small aquatic animals

LIFE EXPECTANCY: 2–3 years

FUN FACT:

Certain species of gobies and shrimp have a very special **symbiotic** relationship. They use their strengths to help each other survive. Gobies have great eyesight, but shrimp do not. So, a shrimp will stay close to a goby, and the goby will alert the shrimp if there is any danger. Gobies are not good at burrowing, so the shrimp will dig a burrow and allow the goby to sleep in the burrow for safety at night.

Gobies have a good sense of direction. They can identify where they are and create a map of their surroundings in their mind.

Eastern Deer Mouse

FOUND IN: Every region of the United States except the Southeast

SIZE: 5–8 inches (12.7–20.3 cm)

DIET: Seeds, nuts, fruit, insects

LIFE EXPECTANCY: 1–3 years

Deer mice have developed many ways to communicate with each other. To send a message to other mice, they use movement, produce special scents, and make squeaking noises.

FIRST TO THE FINISH LINE

Peregrine Falcon

FOUND IN: Across the United States, especially the western states and Alaska
SIZE: Length: 14–20 inches (35.5–50.8 cm); Wingspan: 3.5 feet (1 m)
DIET: Birds and mammals
LIFE EXPECTANCY: 10–15 years

This bird is the fastest animal in the world! The peregrine falcon can dive through the air at 240 mph (386 km/h).

Sailfish

FOUND IN: The eastern coasts of Florida and North Carolina, the Gulf Coast
SIZE: Up to 11 feet (3.3 m)
DIET: Fish and cephalopods
LIFE EXPECTANCY: 4–5 years

FUN FACT:
The sailfish can change the color of its body to confuse its prey or communicate with other sailfish.

Sailfish are the fastest fish in the world. They fold their fins back to move even faster through the water and can swim up to 68 mph (110 km/h).

Eastern Coachwhip

FOUND IN: Southeastern United States

SIZE: 3–7 feet (1–2 m) long

DIET: Small mammals, birds, reptiles

LIFE EXPECTANCY: 6–12 years

The eastern coachwhip can move at 3.5 mph (5.6 km/h), making it one of the fastest snake species in North America.

Pronghorn Antelope

FOUND IN: Western United States

SIZE: 3.5–4.5 feet (1–1.3 m) tall

DIET: Grasses and vegetation

LIFE EXPECTANCY: 10–15 years

Pronghorn antelopes are almost as fast as cheetahs! These antelopes can sprint up to 60 mph (96 km/h). But cheetahs can only sprint for a short period of time. Pronghorn antelopes can maintain a speed of 30 mph (48 km/h) for more than 20 miles (32 km)!

FUN FACT:

Speedy Species Around the World

Mexican free-tailed bat: 100 mph (161 km/h)

Horsefly: 90 mph (144.8 km/h)

Cheetah: 70 mph (112.6 km/h)

Ostrich: 56 mph (90 km/h)

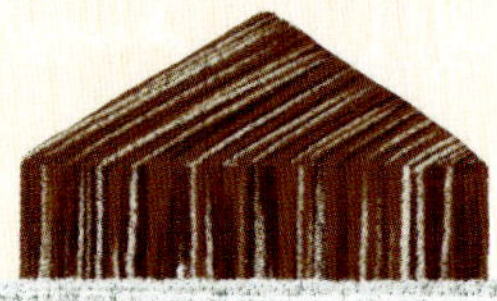

OBSERVING ANIMALS SAFELY

It can be very exciting to see an animal in the wild. But to protect yourself and the animal, it's important to follow some safety tips.

DO:

- Use binoculars or take photos from a safe distance.
- Watch the animal quietly. Loud noises may scare it.
- Leave it in its habitat. If you move it, it may not be able to find its way back home.

DON'T:

- Don't feed a wild animal! It will find food in its natural environment.
- Don't touch the animal or make loud noises; you might scare it!
- Don't disturb its habitat. If you see a nest, hole, or web, look but don't touch. Animals work hard to find and build their habitats!
- Don't approach an animal with its babies! Animals are very protective of their babies.

HOW TO HELP ENDANGERED ANIMALS

Animals are threatened by human activities. When people build houses or stores, forests and other animal habitats are destroyed. Litter, car exhaust, and farms can cause air and water **pollution**.

Because of this, many animals are endangered. Their populations are getting smaller and smaller, and it's up to us to help them!

HERE'S HOW YOU CAN HELP PROTECT ENDANGERED ANIMALS:

- Don't litter!
- When you hike, stay on the trail to avoid disturbing any habitats.
- Plant native flowers and trees to provide habitats for animals.
- Avoid using fertilizers or pesticides in your yard.
- Recycle your waste, and reuse items when you can.
- Buy local, organic food.
- Teach your family and friends about endangered animals and how they can help!

GLOSSARY

adaptation A special body part or behavior that helps an animal survive.

burrow A hole or tunnel in the ground that an animal digs for its habitat.

camouflage A strategy animals use to blend in with their surroundings.

carrion Dead animals that some other animals eat.

endangered When a population of animals is at risk of becoming extinct.

environment The natural world where animals and plants live.

evolve When changes happen over a long period of time so that a species can survive better in its environment.

extinction When a certain type of animal no longer exists on Earth.

gill A body part that fish and some other animals use to breathe underwater.

glacier A large, frozen river of ice that moves very slowly downhill.

habitat A type of environment where an animal or plant lives.

invasive species Animals or plants that are not native to a certain location and can harm the environment.

larvae Young insects that have not yet grown into adult insects.

migrate To move from one place to another.

mimic To look or act like something else.

native species Animals or plants that are naturally found in a certain place.

navigate To get around or move.

nocturnal Active and awake at night instead of during the day.

photoreceptors Special cells in the eyes that help us see light and colors.

pollution An action or material that makes the environment dirty and unhealthy for animals and humans.

population The total number of animals living in a certain place.

predator An animal that hunts other animals for food.

species A group of similar animals.

symbiotic Describes a special relationship where two different living things help each other.

INDEX

A

adaptations, 20–21

B

builders, 30–31
burrows, 13, 14, 20, 30, 41

C

camouflage, 16–17
climate change, 37
communication, 41, 42

E

eating habits and diet, 8–9, 22, 25, 26, 28
endangered species, 3, 10, 12–13, 30
    helping, 45
extinction, 3, 12

F

fastest species, 42–43

G

gill, 4
glaciers, 3, 4
gliding, 13, 38

H

habitats, 44
    extreme, 4–5
    most specific, 26–27
hearing, 7, 15, 21
hibernation, 30

I

invasive species, 14

L

legs, 10–11

M

memory, 40
migration, 17
mimicry, 16

N

nests, 5, 15, 19, 30, 31, 44
nocturnal animals, 11

O

oak savanna, 26
observing animals safely, 44

P

patagium, 13
photoreceptors, 6
"playing possum," 14
pollution, 45
predators/hunters, 34–37
    protection from, 10, 14–15, 16–17, 23, 39

R

rafting, by otters, 40

S

scavengers, 25
sight/eyes, 6, 23, 41
size
    large animals, 29
    tiny animals, 28
smartest species, 40–41
smell, sense of, 22, 23
sounds, 32–33
symbiotic relationships, 41

T

teeth, 22, 30, 31, 34
tools, 40
tortoise and turtle, difference between, 13

V

venom, 23, 36, 37

W

whiskers, 33
wingspan, 18–19

Little Brown Myotis Bat
American Beaver
Saddleback Caterpillar
American Bison (Buffalo)
Ruby-Throated Hummingbird
Raccoon
Sailfish
American Burying Beetle
Northern Idaho Ground Squirrel